THE ALEUTIAN ISLANDS (THE PACIFIC RIM OF FIRE) AND THEIR WILDLIFE

Kerry Atkinson

A Book in the Nature's Wonders Series

AF875209

Every effort has been made in preparing this book to provide accurate and up-to-date information which is accordance with accepted standards and practice at the time of publication. The author and publisher can make no warranties that the information contained herein is totally free from error. The author and publisher disclaim all liability for direct or consequential damages resulting from the use of material contained in this book.

About this book

The painting on the front cover "Tufted puffin and Sealion" is by Kerry Atkinson© All rights reserved. All the text was written by the author. All images are in the public domain, and the reason why they are in the public domain is described beside each respective image

Key words: Aleutian Islands, Pacific rim of fire, travel, puffins, sealions, volcanoes

Acknowledgements

I would like to acknowledge the help that I received from Jim Severino in designing the covers for this book and for uploading it for me to Kindle Direct Publishing.

Other books by Kerry Atkinson

Medical books

1. Graft-versus-Host Disease. Editors, S Burakoff, H J Deeg, J Ferrara, K Atkinson. Marcel Dekker Inc, New York, 1990.
2. Clinical Bone Marrow Transplantation. First edition. Editor, K Atkinson. Cambridge University Press, Cambridge UK, 1994.
3. The BMT Data Book: A Manual for Bone Marrow and Blood Stem Cell Transplantation. K. Atkinson, Cambridge University Press, Cambridge UK, 1998.
4. Clinical Bone Marrow and Blood Stem Cell Transplantation. Second Edition. Editor, K. Atkinson. Cambridge University Press, Cambridge UK, 2000.
5. Clinical Bone Marrow and Blood Stem Cell Transplantation. Third Edition. Eds, K. Atkinson, R Champlin, J Ritz, M Brenner, P Ljungman and W Fibbe. Cambridge University Press, Cambridge UK, 2004.
6. The BMT Data Book. Second edition. R Munker, HM Lazarus and K. Atkinson, Cambridge University Press, Cambridge UK, 2009.

40. Nature's Wonders: Tanzania. Kindle Direct Publishing 2022. e-book and paperback.
41. The Silk Road: Uzbekistan and Kazakhstan – A Book in the Nature's Wonders Series. 2022. Kindle Direct Publishing 2022. e-book and paperback.
42. Madagascar – Land of Lemurs and Chameleons: a book in the Nature's Wonders series 2023. Kindle Direct Publishing 2022. e-book and paperback.
43. South Africa - Kalahari to the Cape - a book in the Nature's Wonders series. Kindle Direct Publishing 2022. e-book and paperback
44. Nature's Wonders – Finland - land of owls and lakes. Kindle Direct Publishing 2023. e-book and paperback
45. Nature's Wonders – Ethiopia. Kindle Direct Publishing 2023. e-book and paperback
46. Nature's Wonders – Kakadu National Park and Central Australia. Kindle Direct Publishing 2023. e-book and paperback
47. Nature's Wonders – Lord Howe Island, Australia. Kindle Direct Publishing 2023. e-book and paperback
48. Norway above the Arctic Circle in March. A book in the Nature's Wonders series. Kindle Direct Publishing 2024. e-book and paperback
49. The Inside Passage and the Pribilof Islands. A book in the Nature's Wonders series. Kindle Direct Publishing 2024. e-book and paperback
50. Orca! A book in the Nature's Wonders series. Kindle Direct Publishing 2024. e-book
51. Watching Humpback Whales from your Sofa, while sipping a Glass of Chardonnay (Australian of course) or…The Lazy Naturalist's Guide to Local Sydney Wildlife. Kindle Direct Publishing 2024. e-book and paperback
52. Magic Moments of an Amateur Naturalist. A book in the Nature's Wonder's series. Kindle Direct Publishing 2024. e-book and paperback
53. Amazing evolution Volume 1. Kindle Direct Publishing 2024. e-book and paperback
54. Nature's Wonders – the USA. Kindle Direct Publishing 2024. e-book and paperback
55. Amazing evolution Volume 2. Kindle Direct Publishing 2024. e-book and paperback

56. South Island, New Zealand. Kindle Direct Publishing 2024. e-book and paperback
57. Deserts of the World and their wildlife. Kindle Direct Publishing 2025. e-book and paperback
58. Mountain ranges of Central Asia and their wildlife. Kindle Direct Publishing 2025. e-book and paperback
59. Brisbane Birds. A book in the Nature's Wonders series. Kindle Direct Publishing 2025. e-book and paperback
60. Lake Baikal and its wildlife. A book in the Nature's Wonders series. Kindle Direct Publishing 2025. e-book and paperback
61. Little known seas of the World and their wildlife. Kindle Direct Publishing 2025. e-book and paperback

Other books

62. Paintings by Kerry Atkinson 2019 – 2021. Kindle Direct Publishing 2021. e-book and paperback.
63. Ferrari Road Cars. Kindle Direct Publishing 2022. e-book and paperback.
64. Paintings by Kerry Atkinson. 2022. Kindle Direct Publishing 2022. e-book and paperback.
65. Samuel Beckett and Bob Dylan – living in a Shakespearean era. Kindle Direct Publishing 2022. e-book and paperback.
66. Paintings and sculptures by Kerry Atkinson 2023. Kindle Direct Publishing 2023. e-book and paperback.
67. Ten influential European paintings. Kindle Direct Publishing 2023. e-book and paperback
68. The Pillars of Rock 'N Roll. Kindle Direct Publishing 2024. e-book and paperback
69. Early 20th Century Western paintings - a smorgasbord of color and radical shapes. Kindle Direct Publishing 2024. e-book
70. Insects and other wildlife included in Dutch flower paintings from the Dutch golden age. Kindle Direct Publishing 2024. e-book and paperback
71. Paintings and sculptures by Kerry Atkinson 2024

Other books

72. Paintings by Kerry Atkinson 2019 – 2021. Kindle Direct Publishing 2021. e-book and paperback.
73. Ferrari Road Cars. Kindle Direct Publishing 2022. e-book and paperback.
74. Paintings by Kerry Atkinson. 2022. Kindle Direct Publishing 2022. e-book and paperback.
75. Samuel Beckett and Bob Dylan – living in a Shakespearean era. Kindle Direct Publishing 2022. e-book and paperback.
76. Paintings and sculptures by Kerry Atkinson 2023. Kindle Direct Publishing 2023. e-book and paperback.
77. Ten influential European paintings. Kindle Direct Publishing 2023. e-book and paperback
78. The Pillars of Rock 'N Roll. Kindle Direct Publishing 2024. e-book and paperback
79. Early 20th Century Western paintings - a smorgasbord of color and radical shapes. Kindle Direct Publishing 2024. e-book
80. Insects and other wildlife included in Dutch flower paintings from the Dutch golden age. Kindle Direct Publishing 2024. e-book and paperback
81. Paintings and sculptures by Kerry Atkinson 2024
82. Cadillacs with Rocket-like Tailfins. Kindle Direct Publishing 2025. e-book and paperback
83. Boy Racer. Kindle Direct Publishing 2025. e-book and paperback

Table of Contents

1. Author's preface

I have visited the Pribilof Islands which are north of the Aleutian Islands, but sadly not the Aleutians themselves. I did fly over them on a commercial flight from Japan to the USA and, looking down on then, molten lava was visible in some of them. How thrilling.

2. The Aleutian Islands and their wildlife

The Aleutian Islands, also called the Aleut Islands, Aleutic Islands, and, before 1867, the Catherine Archipelago, are a chain of 14 main volcanic islands and 55 smaller ones. Catherine refers to Catherine I, Empress of Russia (1684 – 1727), before the USA bought Alaska from Russia in 1867. Catherine was the widow of Peter the Great of Russia

Figure 1. Equestrian portrait of Empress Catherine I

This image is after Georg Cristoph Grooth - Art Museum of Estonia's Digital Archive, Public Domain, https://commons.wikimedia.org/w/index.php?curid=12236364; https://en.wikipedia.org/wiki/Catherine_I_of_Russia#/media/File:Equestrian_portrait_of_Catherine_I.jpg. The author died in 1749, so this work is in the public domain in its country of origin and other countries and areas where the copyright term is the author's life plus 100 years or fewer.This work is in the public domain in the United States because it was published (or registered with the U.S. Copyright Office) before January 1, 1930.

Most of the Aleutian Islands belong to the U.S. state of Alaska, with the archipelago encompassing the Aleutians West Census Area and the Aleutians East Borough. The Commander Islands, located further to the west, belong to the Russian federal subject of Kamchatka Krai, of the Russian Far East

Nearly all of the Aleutians are protected as part of the Alaska Maritime National Wildlife Refuge and the Aleutian Islands Wilderness. The Aleutian Islands were designated a UNESCO biosphere reserve in 1976.

The islands form part of the Aleutian Arc of the Northern Pacific Ocean, and occupy a land area of 6,821 sq mi (17,666 km^2) that extends westward roughly 1,200 mi (1,900 km) from the Alaska Peninsula mainland, in the direction of the Kamchatka Peninsula of Russia. The archipelago acts as a border between the Bering Sea to the north and the Pacific Ocean to the south. The westernmost U.S. island is Attu Island

The islands, with their 57 volcanoes, form the northernmost part of the Pacific Ring of Fire.

Battles and skirmishes took place during the Aleutian Islands campaign of World War II. The Japanese landing and occupations of Kiska and Attu, in June of 1942, were one of only two invasions of the United States in North America during WWII.

Figure 2. Location of the Aleutian Islands.

The Aleutian Islands are highlighted in this image, stretching from Alaska almost to Russia. The two land masses are the Russian Far East/Siberia on the left and Alaska on the right. The Bering Strait separates them and the large island in the middle of the Strait is St Lawrence Island. The Pribilof Islands are in the Bering Sea below St Lawrence Island.

This image is by Aleuten_Lage.png: modified by Carportderivative work: Amirki - This file was derived from: Aleuten Lage.png:, GFDL, https://commons.wikimedia.org/w/index.php?curid=21154712; en.wikipedia.org/wiki/Aleutian_Islands#/media/File:Aleuten_Lag.e-blank.png. Permission is granted to copy, distribute and/or modify this document under the terms of the **GNU Free Documentation License**, Version 1.2 or any later version published by the Free Software Foundation.

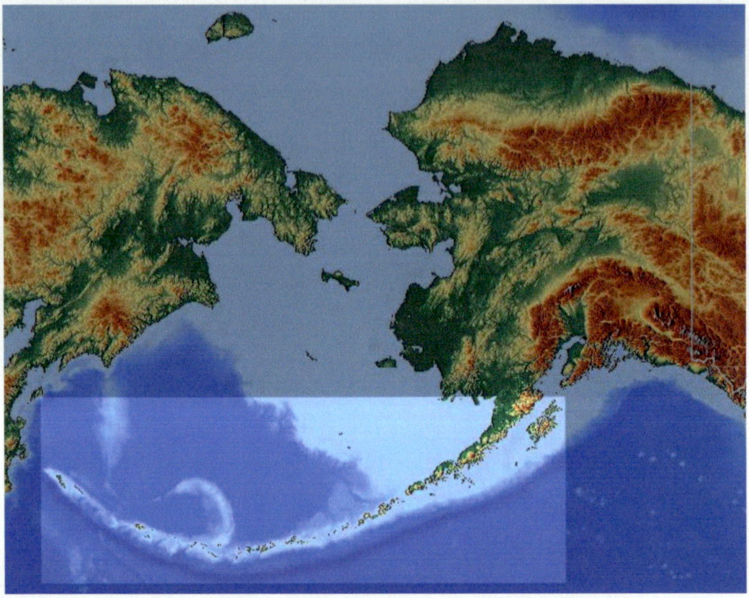

Figure 3. Aleutian Islands on May 15, 2014, by NASA's Aqua satellite

This image is by NASA image courtesy Jeff Schmaltz, LANCE/EOSDIS MODIS Rapid Response Team at NASA GSFC. - http://earthobservatory.nasa.gov/IOTD/view.php?id=83822, Public Domain, https://commons.wikimedia.org/w/index.php?curid=33238408; en.wikipedia.org/wiki/Aleutian_Islands#/media/File:Aleutian_Islands_amo_2014135_lrg.jpg. This file is in the public domain in the United States because it was solely created by NASA.

Figure 4. Active Aleutian volcanoes

This image is by Public Domain, https://commons.wikimedia.org/w/index.php?curid=533060; en.wikipedia.org/wiki/Aleutian_Islands#/media/File:North-Pacific-air-routes.png. This image is in the public domain in the United States because it only contains materials that originally came from the United States Geological Survey, an agency of the United States Department of the Interior.

These North Pacific air routes carry more than 20,000 people and millions of dollars in cargo every day.

The Aleutian Islands include six groups (east to west):

Fox Islands: the main islands are Unimak, Akutan, Unalaska and Umnak

Islands of Four Mountains: the main islands are Yunaska and Chuginadak

Andreanof Islands: the main islands are
Adak, Atka, Amlia, Seguam, Kanaga and Tanaga

Rat Islands: the main islands are Kiska and Amchitka

Near Islands: the main islands are Attu Island, Agattu Island and the Semichi Islands - Alaid, Nizki and Shemya

Commander Islands: the main islands are Bering and Medny

The largest island is Attu - the farthest from the mainland.

The majority of the islands bear evident marks of volcanic origin. These volcanic islands reach heights of 6,200 feet (1,900 m). The coasts are rocky, and the approaches are dangerous, the land rising immediately from the coasts to steep mountains.

Figure 5. Cape Promontory, Cape Lutkes on Unimak Island

This image is by U.S Fish and Wildlife Service - FWS National Digital Library – file, Public Domain, https://commons.wikimedia.org/w/index.php?curid=9739352; en.wikipedia.org/wiki/Aleutian_Islands#/media/File:Cape_Promontory_Cape_Lutkes,_Alaska._Cape_Lutkes_is_part_of_the_Aleutian_Islands.JPG. This image or recording is the work of a U.S. Fish and Wildlife Service employee, taken or made as part of that person's official duties. As a work of the U.S. federal government, the image is in the public domain.

The climate of the islands is oceanic, with moderate and fairly uniform temperatures and heavy rainfall. Fogs are almost constant

The mean annual temperature for Unalaska, the most populated island of the group, is about 38 °F (3 °C), being about 30 °F (−1 °C) in January and about 52 °F (11 °C) in August.

Figure 6. Unalaska Island. The Russian Orthodox church is at the left end of the photograph.

This image is by Wanetta Ayers - http://www.fhwa.dot.gov/byways/photos/65168, Public Domain, https://commons.wikimedia.org/w/index.php?curid=689244; en.wikipedia.org/wiki/Aleutian_Islands#/media/File:UnalaskaAlaska.jpg. This work has been released into the public domain by its author, Wanetta Ayers.

Attu Island

Attu is the Aleutian Island furthest from mainland Alaska. It is right at the end of the archipelago. It is currently uninhabited.

Figure 7. Location of Attu Island, marked by the red dot

This image is by CC BY-SA 3.0, https://commons.wikimedia.org/w/index.php?curid=852303; https://en.wikipedia.org/wiki/Attu_Island#/media/File:AKMap-doton-AttuStation.PNG. Permission is granted to copy, distribute and/or modify this document under the terms of the **GNU Free Documentation License**, Version 1.2 or any later version published by the Free Software Foundation.

Flora

With the exception of some stunted willows, the vast majority of the chain is devoid of native trees. On some of the islands, such as Adak and Amaknak, there are a few coniferous trees growing, remnants of the Russian period. While tall trees grow in many cold climates, Aleutian conifers — some estimated to be two hundred years old — rarely reach a height of even 10 feet (3 m), and many of them are less than 5 feet (1.5 m) tall. This is because the islands experience such strong winds

Instead of trees, the islands are covered with a luxuriant, dense growth of herbage and shrubs, including Crowberry, Bluejoint, grasses, sedges, and many flowering plants. There are areas of peat bog near the coasts. Endemic plants include the endangered Aleutian shield fern.

Figure 8. Bluejoint (aka bluejoint reedgrass, marsh reedgrass, Canadian reedgrass, meadow pinegrass, Canada bluejoint and marsh pinegrass). (*Calamagrostis canadensis)*

This image is by Robert H. Mohlenbrock @ USDA-NRCS PLANTS Database / USDA SCS, Public Domain, https://commons.wikimedia.org/w/index.php?curid=16387695; en.wikipedia.org/wiki/Calamagrostis_canadensis#/media/File:Calamagrostis-canadensis1we.jpg. This image or file is a work of a United States Department of Agriculture employee, taken or made as part of that person's official duties. As a work of the U.S. federal government, the image is in the public domain.

Fauna

Orcas and Steller's sealions occur in the Aleutians.

Figure 9. Orcas breaching near Unimak Island in the eastern Aleutian Islands.

This image is by Robert Pittman - NOAA (http://www.afsc.noaa.gov/Quarterly/amj2005/divrptsNMML3.htm]), Public Domain, https://commons.wikimedia.org/w/index.php?curid=1433661; en.wikipedia.org/wiki/Orca#/media/File:Killerwhales_jumping.jpg. This image is in the public domain because it contains materials that originally came from the U.S. National Oceanic and Atmospheric Administration, taken or made as part of an employee's official duties.

Figure 10. Steller's sea lions on Amak Island

This image is by Bell, Kevin - U.S. Fish and Wildlife Service Public Domain, https://commons.wikimedia.org/w/index.php?curid=11516800; en.wikipedia.org/wiki/Aleutian_Islands#/media/File:Amak_Island,_Steller's_Sea_Lion_haul_out.jpg. This image or recording is the work of a U.S. Fish and Wildlife Service employee, taken or made as part of that person's official duties. As a work of the U.S. federal government, the image is in the public domain.

Sea otters occur along the coasts of many of the Aleutian Islands. Their presence encourages the growth of kelp forests, as the otters control sea urchin populations. Large populations of sea urchins can create urchin barrens by clearing away kelp stands.

Figures 11 – 13. Sea otters

The Sea otter's thick fur makes its body appear plumper on land than in the water.

This image is by Regional Director-External Affairs, U.S. Fish and Wildlife Service - AK/RO/00175, Public Domain, https://commons.wikimedia.org/w/index.php?curid=2072793; en.wikipedia.org/wiki/Sea_otter#/media/File:Seaotterrocks.jpg. This image or recording is the work of a U.S. Fish and Wildlife Service employee, taken or made as part of that person's official duties. As a work of the U.S. federal government, the image is in the public domain.

A sea otter in a Kelp forest

This image is by Robert Schwemmer/NOAA - Chumash Heritage National Marine Sanctuary, Public Domain, https://commons.wikimedia.org/w/index.php?curid=154811312; en.wikipedia.org/wiki/Sea_otter#/media/File:Sea_otter_kelp_forest_Chumash_Heritage_National_Marine_Sanctuary.PNG. This image is in the public domain because it contains materials that originally came from the U.S. National Oceanic and Atmospheric Administration, taken or made as part of an employee's official duties.

A pair of Sea otters

This image is by Ed Bowlby, NOAA research co-ordinator - http://www.marinesanctuariesmedia.com/product.php?product_id=104, Public Domain, https://commons.wikimedia.org/w/index.php?curid=3407502 ; en.wikipedia.org/wiki/Sea_otter#/media/File:Sea_otter_pair2.jpg. This image is in the public domain because it contains materials that originally came from the U.S. National Oceanic and Atmospheric Administration, taken or made as part of an employee's official duties.

Figures 14 – 17. Kelp forest

This image is by Kip Evans - Published by National Oceanic and Atmospheric Administration at http://www.photolib.noaa.gov/nurp/nur03505. Public Domain, https://commons.wikimedia.org/w/index.php?curid=78894; en.wikipedia.org/wiki/Kelp_forest#/media/File:Kelp_forest.jpg. This image is in the public domain because it contains materials that originally came from the U.S. National Oceanic and Atmospheric Administration, taken or made as part of an employee's official duties.

This image is by Robert Schwemmer/NOAA - Chumash Heritage National Marine Sanctuary, Public Domain, https://commons.wikimedia.org/w/index.php?curid=154811082; en.wikipedia.org/wiki/Kelp_forest#/media/File:Kelp_forest_Cojo_Anchorage.PNG. This image is in the public domain because it contains materials that originally came from the U.S. National Oceanic and Atmospheric Administration, taken or made as part of an employee's official duties.

Giant kelp uses gas-filled floats to keep the thallus suspended, allowing the kelp blades near the ocean surface to capture light for photosynthesis

A thallus is a plant body that is not differentiated into stem and leaves and lacks true roots and a vascular system.

This image is by Public Domain, https://commons.wikimedia.org/w/index.php?curid=32046; en.wikipedia.org/wiki/Kelp_forest#/media/File:Giantkelp2_300.jpg. This image is in the public domain because it contains materials that originally came from the U.S. National Oceanic and Atmospheric Administration, taken or made as part of an employee's official duties.

This image is by Fred Hsu (Wikipedia:User:Fredhsu on en.wikipedia) - Photo taken and uploaded by user, CC BY-SA 3.0, https://commons.wikimedia.org/w/index.php?curid=1559678; en.wikipedia.org/wiki/Kelp_forest#/media/File:Rockfish_around_kelp_Monterey_Bay_Aquarium.jpg. Permission is granted to copy, distribute and/or modify this document under the terms of the GNU Free Documentation License, Version 1.2 or any later version published by the Free Software Foundation.

Figure 18. Sea urchin

Sea urchins, such as this Purple sea urchin, can damage Kelp forests by chewing through Kelp holdfasts (anchors).

This image is by Public Domain, https://commons.wikimedia.org/w/index.php?curid=32050; en.wikipedia.org/wiki/Kelp_forest#/media/File:Seaurchin_300.jpg. This image is in the public domain because it contains materials that originally came from the U.S. National Oceanic and Atmospheric Administration, taken or made as part of an employee's official duties.

Figure 19. A Jeweled top snail (*Calliostoma annulatum*) grazing on a blade of giant kelp

This image is by Steve Lonhart (SIMoN / MBNMS) - http://www.mbnms-simon.org/other/photos/photo_info.php?photoID=2163, Public Domain, https://commons.wikimedia.org/w/index.php?curid=2054102; en.wikipedia.org/wiki/Kelp_forest#/media/File:Calliostoma_annulatum.jpg. This image is in the public domain because it contains materials that originally came from the U.S. National Oceanic and Atmospheric Administration, taken or made as part of an employee's official duties.

The Aleutians are also home to many large colonies of seabirds. Buldir Island has 21 breeding seabird species, including the Red-legged kittiwake which is endemic to the Bering Sea. Red-legged kittiwakes are resident on the Aleutian Islands. Large seabird colonies are also present at Kiska, Gareloi, Semisopochnoi, Bogoslof, and others.

Figures 20 - 24. Red-legged kittiwakes

This image is by Unknown author - U.S. Fish and Wildlife Service, Public Domain, https://commons.wikimedia.org/w/index.php?curid=283912; https://en.wikipedia.org/wiki/Red-legged_kittiwake#/media/File:Rlegkittiwake6.jpg. This image or recording is the work of a U.S. Fish and Wildlife Service employee, taken or made as part of that person's official duties. As a work of the U.S. federal government, the image is in the public domain.

This image is by Dean Kildaw, USFWS - US Fish and Wildlife Service, Public Domain, https://commons.wikimedia.org/w/index.php?curid=707360 https://en.wikipedia.org/wiki/Red-legged_kittiwake#/media/File:Rissa_brevirostris5.jpg. This image or recording is the work of a U.S. Fish and Wildlife Service employee, taken or made as part of that person's official duties. As a work of the U.S. federal government, the image is in the public domain.

Red-legged Kittiwakes nesting

This image is by Josh Keaton NOAA/NMFS/AKRO/SFD. - NOAA Photo Library: anim1693, Public Domain, https://commons.wikimedia.org/w/index.php?curid=17985477; https://en.wikipedia.org/wiki/Red-legged_kittiwake#/media/File:Red-legged_Kittiwake_nests,_Alaska_-_NOAA.jpg. This image is in the public domain because it contains materials that originally came from the U.S. National Oceanic and Atmospheric Administration, taken or made as part of an employee's official duties.

A Red-legged Kittiwake chick

This image is by Public Domain, https://commons.wikimedia.org/w/index.php?curid=707371; https://en.wikipedia.org/wiki/Red-legged_kittiwake#/media/File:Rissa_brevirostris7.jpg. This image or recording is the work of a U.S. Fish and Wildlife Service employee, taken or made as part of that person's official duties. As a work of the U.S. federal government, the image is in the public domain.

A colony of Red-legged kittiwakes

This image is by Momatiuk and John Eastcott, USFWS - US Fish and Wildlife Service, Public Domain, https://commons.wikimedia.org/w/index.php?curid=707379; en.wikipedia.org/wiki/Red-legged_kittiwake#/media/File:Red-legged_Kittiwake_colony,_St._George_Island,_Alaska.jpg. This image or recording is the work of a U.S. Fish and Wildlife Service employee, taken or made as part of that person's official duties. As a work of the U.S. federal government, the image is in the public domain.

The Aleutian tern

The eponymous Aleutian Tern also breeds on the Aleutian Islands but migrates in winter on a long west-east migration to the tropical seas off Malaysia and Thailand – a one-way distance of some 9,713 kilometers (6,000 miles).

Figure X. Aleutian tern

This image is by usgs.gov - https://alaska.usgs.gov/science/biology/avian_influenza/species/species.php?code=ALTE, Public Domain, https://commons.wikimedia.org/w/index.php?curid=23312083; en.wikipedia.org/wiki/Aleutian_tern#/media/File:Aleutian_Tern.jpg. This work is in the public domain in the United States because it is a work prepared by an officer or employee of the United States Government as part of that person's official duties under the terms of Title 17, Chapter 1, Section 105 of the US Code.

It often nests alongside Arctic terns but is less defensive of its nest and young against predators such as Arctic jaegers (skuas in Europe) than Arctic terns, which fiercely defend their nests.

Figure 26. Arctic (or Parasitic) jaeger

This image is by Alaska Region U.S. Fish & Wildlife Service - https://www.flickr.com/photos/usfws_alaska/51358219005/, Public Domain, https://commons.wikimedia.org/w/index.php?curid=115267867; en.wikipedia.org/wiki/Parasitic_jaeger#/media/File:2021-07-11_Parasitic_jaeger_-_51358219005.jpg. This image or recording is the work of a U.S. Fish and Wildlife Service employee, taken or made as part of that person's official duties. As a work of the U.S. federal government, the image is in the public domain.

Other resident birds of the Aleutians include the Red-faced cormorant, Snow bunting and Common loon (known as the Great northern diver in Europe)

Figure 27. Red-faced cormorant

This image is by Lisa Hupp/USFWS, Public Domain, By Lisa Hupp/USFWS. - en.wikipedia.org/wiki/Red-faced_cormorant#/media/File:Red-faced_Cormorant_on_St._Paul_Island_by_Lisa_Hupp_USFWS.jpghttpsThis work is in the public domain in the United States because it is a work prepared by an officer or employee of the United States Government as part of that person's official duties under the terms of Title 17, Chapter 1, Section 105 of the US Code.

Figure 28. Snow bunting

This image is by Alaska Region U.S. Fish & Wildlife Service - https://www.flickr.com/photos/usfws_alaska/52828320918, Public Domain, https://commons.wikimedia.org/w/index.php?curid=158433688. This image or recording is the work of a U.S. Fish and Wildlife Service employee, taken or made as part of that person's official duties. As a work of the U.S. federal government, the image is in the public domain.

Figure 29. Common loon fishing

This image is by ALAN SCHMIERER - https://en.wikipedia.org/wiki/Common_loon#/media/File:093_-_COMMON_LOON_(12-17-11)_moss_landing,_ca_(2)_(8720312789).jpghttps://www.flickr.com/photos/sloalan/8720312789/, CC0, https://commons.wikimedia.org/w/index.php?curid=57032295;. This file is made available under the Creative Commons CC0 1.0 Universal Public Domain Dedication.

hile Red-legged kittiwakes are one of a number of breeding birds on the Aleutian Islands, Astonishinglythe avifauna of the Aleutian Islands includes 376 species according to Gibson and Byrd (2007), Bird Checklists of the World (Avibase), and the American Ornithological Society (AOS) as of March 2021.

Of these, 44 (12%) are year-round residents and breeders, 26 (7%) migrate to the Aleutians to breed, 18 (5%) migrate to the Aleutians to winter, 6 (2%) are non-breeding summer residents, 37 (10%) are annual through-migrants, and 196 (53%) are vagrants (aka accidentals) of less-than-annual occurrence. Several of the vagrants have only a single record. The status of the remaining 42 is not noted by either Gibson and Byrd or Avibase.

The following terms are used to denote the annual and seasonal status of each species reported by Gibson and Byrd (2007), or in the case of "Accidental", noted as such for species added from Avibase:

Accidental – one or two records

Casual – recorded in less than 30% of years in the appropriate season, but in at least three calendar years

Intermittent – recorded in 30% or more years in the appropriate season, but not annually

Migrant – annual through-migrant in spring or fall

Resident – substantial numbers present throughout the year

Summer – migrates to the Aleutians to breed or to summer offshore

Winter – migrates to the Aleutians to winter

Annual breeders are designated with an asterisk (*), as in resident* or summer*

The complete list is as follows:

Ducks, geese, and waterfowl

Order: Anseriformes   Family: Anatidae

Emperor goose, Anser canagica (winter)

Lesser white-fronted goose, Anser erythropus (accidental)

Taiga bean-goose, Anser fabalis (intermittent)

Tundra bean-goose, Anser serrirostris

Brant, Branta bernicla (migrant)

Cackling goose, Branta hutchinsii (summer*)

Canada goose, Branta canadensis (accidental)

Trumpeter swan, Cygnus buccinator

Tundra swan, Cygnus columbianus (resident*)

Whooper swan, Cygnus cygnus (winter)

Baikal teal, Sibirionetta formosa (casual)

Garganey, Spatula querquendula (intermittent)

Figure 30. Emperor geese at Adak Island

This image is by Deines, F - U.S. Fish & Wildlife Service, Public Domain, https://commons.wikimedia.org/w/index.php?curid=11517116; en.wikipedia.org/wiki/Emperor_goose#/media/File:Emperor_Geese_at_Adak_Island_ Clam_Lagoon.jpg. This image or recording is the work of a U.S. Fish and Wildlife Service employee, taken or made as part of that person's official duties. As a work of the U.S. federal government, the image is in the public domain

Snow goose, Anser caerulescens (casual)

Ross's goose, Anser rossii (accidental)

Greater white-fronted goose, Anser albifrons (migrant)

Blue-winged teal, Spatula discors (accidental)

Northern shoveler, Spatula clypeata (migrant)

Gadwall, Mareca strepera (winter)

Falcated duck, Mareca falcata (intermittent)

Eurasian wigeon, Mareca penelope (migrant)

American wigeon, Mareca americana (intermittent)

Eastern spot-billed duck, Anas zonorhyncha (casual)

Mallard, Anas platyrhynchos (resident*)

Northern pintail, Anas acuta (resident*)

Green-winged teal, Anas crecca (resident*)

Canvasback, Aythya valisneria (casual or intermittent)

Redhead, Aythya americana (accidental)

Common pochard, Aythya ferina (intermittent)

Ring-necked duck, Aythya collaris (casual)

Tufted duck, Aythya fuligula (migrant)

Greater scaup, Aythya marila (resident*)

Lesser scaup, Aythya affinis (casual or intermittent)

Steller's eider, Polysticta stelleri (winter)

Spectacled eider, Somateria fischeri (accidental)

King eider, Somateria spectabilis (winter)

Common eider, Somateria mollissima (resident*)

Harlequin duck, Histrionicus histrionicus (resident)

Figure 31. A pair of Harlequin ducks, hen (left) and drake (right)

This image is by IOHANNVSVERVS - Own work, CC0, https://commons.wikimedia.org/w/index.php?curid=157730620; en.wikipedia.org/wiki/Harlequin_duck#/media/File:Harlequin_ducks.jpg. This file is made available under the Creative Commons CC0 1.0 Universal Public Domain Dedication.

Surf scoter, Melanitta perspicillata (casual or intermittent)

White-winged scoter, Melanitta deglandi (winter)

Stejneger's Scoter, Melanitta stejnegeri[2]

Black scoter, Melanitta amerivcana (winter)

Long-tailed duck, Clangula hyemalis (winter)

Bufflehead, Bucephala albeola (winter)

Common goldeneye, Bucephala clangula (winter)

Barrow's goldeneye, Bucephala islandica (winter)

Figure X. Male Barrow's Goldeneye

Smew, Mergellus albellus (winter)

Hooded merganser, Lophodytes cucullatus (casual)

Common merganser, Mergus merganser (resident)

Figure 32. A male Barrow's Goldeneye

This image is by Public Domain, https://commons.wikimedia.org/w/index.php?curid=1526406; en.wikipedia.org/wiki/Barrow%27s_goldeneye#/media/File:Barrows.goldeneye.male.010107.arp.jpg. This work has been released into the public domain by its author, Arpingstone.

Red-breasted merganser, Mergus serrator (resident*)

Pheasants, grouse, and allies

Order: Galliformes   Family: Phasianidae

Willow ptarmigan, Lagopus lagopus (resident*)

Rock ptarmigan, Lagopus mutus (resident*)

Grebes

Order: Podicipediformes   Family: Podicipedidae

Horned grebe, Podiceps auritus (winter)

Red-necked grebe, Podiceps grisegena (winter)

Western grebe, Aechmophorus occidentalis

Pigeons and doves

Order: Columbiformes   Family: Columbidae

Rock pigeon, Columba livia (accidental)

Oriental turtle-dove, Streptopelia orientlalis (casual or accidental)

Cuckoos

Order: Cuculiformes   Family: Cuculidae

Common cuckoo, Cuculus canorus (intermittent)

Oriental cuckoo, Cuculus optatus (casual)

Nightjars

Order: Caprimulgiformes   Family: Caprimulgidae

Grey nightjar, Caprimulgus jotaka (accidental)

Swifts

Order: Apodiformes   Family: Apodidae

Chimney swift, Chaetura pelagica (accidental)

White-throated needletail, Hirundapus caudacutus (casual)

Common swift, Apus apus (accidental)

Fork-tailed swift, Apus pacificus (casual)

Hummingbirds

Order: Apodiformes   Family: Trochilidae

Rufous hummingbird, Selasphorus rufus (accidental)[2]

Rails, gallinules, and coots

Order: Gruiformes   Family: Rallidae

Sora, Porzana carolina (accidental)

Common moorhen, Gallinula chloropus (accidental)

Eurasian Coot, Fulica atra (accidental)

American coot, Fulica americana (accidental)

Cranes

Order: Gruiformes   Family: Gruidae

Sandhill crane, Antigone canadensis (summer*)

Stilts and avocets

Order: Charadriiformes   Family: Recurvirostridae

Black-winged stilt, Himantopus himantopus (casual or accidental)

Oystercatchers

Order: Charadriiformes   Family: Haematopodidae

Eurasian oystercatcher, Haematopus ostralegus (accidental)

Black oystercatcher, Haematopus bachmani (resident*)

Plovers and lapwings

Order: Charadriiformes   Family: Charadriidae

Northern lapwing, Vanellus vanellus (accidental)

Black-bellied plover, Pluvialis squatarola (migrant)

European golden-plover, Pluvialis apricaria (accidental)

American golden-plover, Pluvialis dominica (accidental)

Pacific golden-plover, Pluvialis fulva (migrant)

Eurasian dotterel, Charadrius morinellus (casual)

Killdeer, Charadrius vociferus (accidental)

Common ringed plover, Charadrius hiaticula (casual)

Semipalmated plover, Charadrius semipalmatus (summer*)

Little ringed plover, Charadrius dubius (casual)

Lesser sand-plover, Charadrius mongolus (migrant)

Sandpipers and allies

Order: Charadriiformes Family: Scolopacidae

Upland sandpiper, Bartramia longicauda (accidental)

Bristle-thighed curlew, Numenius tahitiensis (intermittent)

Whimbrel, Numenius phaeopus (migrant)

Eskimo curlew, Numenius borealis (accidental, possibly extinct)

Far Eastern curlew, Numenius madagascariensis (intermittent)

Bar-tailed godwit, Limosa lapponica (migrant)

Black-tailed godwit, Limosa limosa (intermittent)

Hudsonian godwit, Limosa haemastica (accidental)

Marbled godwit, Limosa fedoa

Ruddy turnstone, Arenaria interpres (migrant)

Black turnstone, Arenaria melanocephala

Great knot, Calidris tenuirostris (casual)

Red knot, Calidris canutus (casual)

Surfbird, Calidris virgata

Ruff, Calidris pugnax (migrant)

Broad-billed sandpiper, Calidris falcinellus (casual)

Sharp-tailed sandpiper, Calidris acuminata (migrant)

Stilt sandpiper, Calidris himantopus (accidental)

Curlew sandpiper, Calidris ferruginea (casual)

Temminck's stint, Calidris temminckii (intermittent)

Long-toed stint, Calidris subminuta (migrant)

Spoon-billed sandpiper, Calidris pygmeus (casual or accidental)

Red-necked stint, Calidris ruficollis (intermittent)

Sanderling, Calidris alba (winter)

Dunlin, Calidris alpina (migrant)

Rock sandpiper, Calidris ptilocnemis (resident*)

Baird's sandpiper, Calidris bairdii (intermittent)

Little stint, Calidris minuta (casual)

Least sandpiper, Calidris minutilla (summer*)

White-rumped sandpiper, Calidris fuscicollis (accidental)

Figure 33. A well camouflaged White-rumped sandpiper

This image is by Tim Bowman, USFWS - http://images.fws.gov/default.cfm?fuseaction=records.display&CFID=3989350&CFTOKEN=51760576&id=E50138FC-65BF-03E7-21F6F1B629818396, Public Domain, https://commons.wikimedia.org/w/index.php?curid=326857;

en.wikipedia.org/wiki/White-rumped_sandpiper#/media/File:White-rumped_Sandpiper.jpg. This image or recording is the work of a U.S. Fish and Wildlife Service employee, taken or made as part of that person's official duties. As a work of the U.S. federal government, the image is in the public domain.

Buff-breasted sandpiper, Calidris subruficollis (casual)

Pectoral sandpiper, Calidris melanotos (migrant)

Semipalmated sandpiper, Calidris pusilla (casual)

Western sandpiper, Calidris mauri (intermittent)

Short-billed dowitcher, Limnodromus griseus (casual)

Long-billed dowitcher, Limnodromus scolopaceus (migrant)

Jack snipe, Lymnocryptes minimus (accidental)

Solitary snipe, Gallinago solitaria (accidental)[2]

Pin-tailed snipe, Gallinago stenura (casual or accidental)

Common snipe, Gallinago gallinago (migrant)

Wilson's snipe, Gallinago delicata (summer*)

Terek sandpiper, Xenus cinereus (casual or intermittent)

Common sandpiper, Actitis hypoleucos (migrant)

Spotted sandpiper, Actitis macularia (accidental)

Green sandpiper, Tringa ochropus (casual)

Solitary sandpiper, Tringa solitaria (accidental)

Gray-tailed tattler, Tringa brevipes (migrant)

Wandering tattler, Tringa incana (migrant)

Lesser yellowlegs, Tringa flavipes (casual)

Spotted redshank, Tringa erythropus (casual)

Common greenshank, Tringa nebularia (migrant)

Greater yellowlegs, Tringa melanoleuca (casual)

Wood sandpiper, Tringa glareola (migrant)

Marsh sandpiper, Tringa stagnatilis (casual)

Wilson's phalarope, Phalaropus tricolor (accidental)

Red-necked phalarope, Phalaropus lobatus (summer*)

Red phalarope, Phalaropus fulicarius (migrant)

Pratincoles and coursers

Order: Charadriiformes   Family: Glareolidae

Oriental pratincole, Glareola maldivarum (accidental)

Skuas and jaegers

Order: Charadriiformes   Family: Stercorariidae

South polar skua, Stercorarius maccormicki (casual)

Pomarine jaeger, Stercorarius pomarinus (migrant)

Parasitic jaeger, Stercorarius parasiticus (summer*)

Long-tailed jaeger, Stercorarius longicaudus (migrant)

Auks, murres, and puffins

Order: Charadriiformes   Family: Alcidae

Dovekie, Alle alle (casual)

Common murre, Uria aalge (resident*)

Thick-billed murre, Uria lomvia (resident*)

Black guillemot, Cepphus grylle

Pigeon guillemot, Cepphus columba (resident*)

Long-billed murrelet, Brachyramphus perdix (accidental)

Marbled murrelet, Brachyramphus marmoratus (resident*)

Kittlitz's murrelet, Brachyramphus brevirostris (summer*)

Ancient murrelet, Synthliboramphus antiquus (resident*)

Cassin's auklet, Ptychoramphus aleuticus (summer*)

Parakeet auklet, Aethia psittacula (summer*)

Least auklet, Aethia pusilla (resident*)

Whiskered auklet, Aethia pygmaea (resident*)

Crested auklet, Aethia cristatella (resident*)

Rhinoceros auklet, Cerorhinca monocerata (casual)

Horned puffin, Fratercula corniculata (summer*)

Tufted puffin, Fratercula cirrhata (resident*)

Gulls, terns, and skimmers

Order: Charadriiformes   Family: Laridae

Black-legged kittiwake, Rissa tridactyla (resident*)

Red-legged kittiwake, Rissa brevirostris (summer*)

Ivory gull, Pagophila eburnea (accidental)

Sabine's gull, Xema sabini (migrant)

Bonaparte's gull, Chroicocephalus philadelphia (accidental)

Black-headed gull, Chroicocephalus ridibundus (migrant)

Ross's gull, Rhodostethia rosea (casual or accidental)

Franklin's gull, Leucophaeus pipixcan (accidental)

Pallas's gull, Ichthyaetus ichthyaetus (accidental)

Black-tailed gull, Larus crassirostris (casual)

Common gull, Larus canus

Short-billed gull, Larus brachyrhynchus

Ring-billed gull, Larus delawarensis (accidental)

Herring gull, Larus argentatus (migrant)

Iceland gull, Larus glaucoides (casual)

Lesser black-backed gull, Larus fuscus (accidental)

Slaty-backed gull, Larus schistisagus (migrant)

Glaucous-winged gull, Larus glaucescens (resident*)

Glaucous gull, Larus hyperboreus (winter)

Sooty tern, Onychoprion fuscatus (accidental)

Aleutian tern, Onychoprion aleuticus (summer*)

Little/Least tern, Sternula albifrons/antillarum (accidental)

Caspian tern, Hydroprogne caspia (accidental)

White-winged tern, Chlidonias leucopterus (accidental)

Common tern, Sterna hirundo (intermittent)

Arctic tern, Sterna paradisaea (summer*)

Loons

Order: Gaviiformes   Family: Gaviidae

Red-throated loon, Gavia stellata (summer*)

Arctic loon, Gavia arctica (migrant)

Pacific loon, Gavia pacifica (migrant)

Common loon, Gavia immer (resident*)

Yellow-billed loon, Gavia adamsii (winter)

Albatrosses

Order: Procellariiformes   Family: Diomedeidae

Salvin's albatross, Thalassarche salvini (accidental)

Laysan albatross, Phoebastria immutabilis (summer)

Black-footed albatross, Phoebastria nigripes (summer)

Short-tailed albatross, Phoebastria albatrus (summer)

Northern storm-petrels

Order: Procellariiformes   Family: Hydrobatidae

Fork-tailed storm-petrel, Hydrobates furcatus (resident*)

Figure 34. The Fork-tailed storm petrel frequents the open ocean and only returns to land to breed.

This image is by Stanley, Carla, Public Domain, https://commons.wikimedia.org/w/index.php?curid=871389; en.wikipedia.org/wiki/Fork-tailed_storm_petrel#/media/File:Fork_tail.jpg. This image or recording is the work of a U.S. Fish and Wildlife Service employee, taken or made as part of that person's official duties. As a work of the U.S. federal government, the image is in the public domain.

Leach's storm-petrel, Hydrobates leucorhous (summer*)

Shearwaters and petrels

Order: Procellariiformes   Family: Procellariidae

Northern fulmar, Fulmarus glacialis (resident*)

Providence petrel, Pterodroma solandri (accidental)

Mottled petrel, Pterodroma inexpectata (summer)

Cook's petrel, Pterodroma cookii (casual or accidental)

Buller's shearwater, Ardenna bulleri (accidental)

Short-tailed shearwater, Ardenna tenuirostris (summer)

Sooty shearwater, Ardenna griseus (summer)

Pink-footed shearwater, Ardenna creatopus (accidental)

Flesh-footed shearwater, Ardenna carneipes (accidental)

Frigatebirds

Order: Suliformes   Family: Fregatidae

Magnificent frigatebird, Fregata magnificens (accidental)

Boobies and gannets

Order: Suliformes   Family: Sulidae

Brown booby, Sula leucogaster (accidental)

Cormorants and shags

Order: Suliformes   Family: Phalacrocoracidae

Red-faced cormorant, Urile urile (resident*)

Pelagic cormorant, Urile pelagicus (resident*)

Double-crested cormorant, Nannopterum auritum (resident*)

Herons, egrets, and bitterns

Order: Pelecaniformes   Family: Ardeidae

Yellow bittern, Ixobrychus sinensis (accidental)

Great blue heron, Ardea herodias (accidental)

Gray heron, Ardea cinerea (accidental)

Great egret, Ardea alba (casual)

Egretta intermedia (accidental)

Chinese egret, Egretta eulophotes (accidental)

Little egret, Egretta garzetta (accidental)

Cattle egret, Bubulcus ibis (accidental)

Chinese pond-heron, Ardeola bacchus (accidental)

Black-crowned night-heron, Nycticorax nycticorax (casual)

Osprey

Order: Accipitriformes   Family: Pandionidae

Osprey, Pandion haliaetus (accidental)

Hawks, eagles, and kites

Order: Accipitriformes   Family: Accipitridae

Golden eagle, Aquila chryaetos (resident*)

Northern harrier, Circus hudsonius (casual)

Sharp-shinned hawk, Accipiter striatus (accidental)

Eurasian goshawk, Accipiter gentiles (accidental)

American goshawk, Accipiter atricapillus (accidental)

Black kite, Milvus migrans (accidental)[2]

Bald eagle, Haliaeetus leucocephalus (resident*)

White-tailed eagle, Haliaeetus albicilla (resident*)

Steller's sea eagle, Haliaeetus pelagicus (casual)

Red-tailed hawk, Buteo jamaicensis (accidental)

Rough-legged hawk, Buteo lagopus (summer*)

Long-legged buzzard, Buteo rufinus (accidental)[2]

Owls

Order: Strigiformes   Family: Strigidae

Oriental scops-owl, Otus sunia (accidental)

Great Horned owl, Bubo virginianus (accidental)

Snowy owl, Bubo scandiacus (resident*)

Long-eared owl, Asio otus (accidental)[2]

Short-eared owl, Aso flammeus (summer*)

Boreal owl, Aegolius funereus (accidental)

Northern saw-whet owl, Aegolius acadicus (accidental)

Northern boobook, Ninox scutulata (accidental)

Kingfishers

Order: Coraciiformes   Family: Alcedinidae

Belted kingfisher, Megaceryle alcyon (resident*)

Woodpeckers

Order: Piciformes   Family: Picidae

Black-backed woodpecker, Picoides arcticus (accidental)

Great spotted woodpecker, Dendrocopos major (casual)

Downy woodpecker, Dryobates pubescens (accidental)

Northern flicker, Colaptes auratus (accidental)

Falcons and caracaras

Order: Falconiformes   Family: Falconidae

Eurasian kestrel, Falco tinnunculus (casual)

Merlin, Falco columbarius (casual or intermittent)

Eurasian hobby, Falco subbuteo (casual)

Gyrfalcon, Falco rusticolus (resident*)

Peregrine falcon, Falco peregrinus (resident*)

Tyrant flycatchers

Order: Passeriformes   Family: Tyrannidae

Eastern kingbird, Tyrannus tyrannus (accidental)

Olive-sided flycatcher, Contopus cooperi (accidental)

Western wood-pewee, Contopus sordidulus (accidental)

Yellow-bellied flycatcher, Empidonax flaviventris (accidental)

Alder flycatcher, Empidonax alnorum (accidental)

Say's phoebe, Sayornis saya (accidental)

Vireos, shrike-babblers, and erpornis

Order: Passeriformes   Family: Vireonidae

Warbling vireo, Vireo gilvus (accidental)

Shrikes

Order: Passeriformes   Family: Laniidae

Brown shrike, Lanius cristatus (accidental)

Northern shrike, Lanius borealis (casual or intermittent)

Crows, jays, and magpies

Order: Passeriformes   Family: Corvidae

Black-billed magpie, Pica hudsonia (resident*)

American crow, Corvus brachyrhynchos

Common raven, Corvus corax (resident*)

Tits, chickadees, and titmice

Order: Passeriformes   Family: Paridae

Black-capped chickadee, Poecile atricapillus (resident*)

Larks

Order: Passeriformes   Family: Alaudidae

Eurasian skylark, Alauda arvensis (intermittent)

Horned lark, Eremophila alpestris (casual)

Grassbirds and allies

Order: Passeriformes   Family: Locustellidae

Middendorff's grasshopper warbler, Helopsaltes ochotensis (casual)

Lanceolated warbler, Locustella lanceolata (casual or accidental)

Swallows

Order: Passeriformes   Family: Hirundinidae

Bank swallow, Riparia riparia (summer*)

Tree swallow, Tachycineta bicolor (casual or intermittent)

Violet-green swallow, Tachycineta thalassina (casual or intermittent)

Northern rough-winged swallow, Stelgidopteryx serripennis (accidental)[2]

Purple martin, Progne subis (accidental)

Barn swallow, Hirundo rustica (casual)

Common house-martin, Delichon urbicum (accidental)[2]

Cliff swallow, Petrochelidon pyrrhonota (casual)

Leaf warblers

Order: Passeriformes   Family: Phylloscopidae

Willow warbler, Phylloscopus trochilus (accidental)

Common chiffchaff, Phylloscopus collybita (accidental)

Wood warbler, Phylloscopus sibilatrix (accidental)

Dusky warbler, Phylloscopus fuscatus (casual)

Yellow-browed warbler, Phylloscopus inornatus (accidental)

Arctic warbler, Phylloscopus borealis (intermittent)

Kamchatka leaf warbler, Phylloscopus examinandus (accidental)[2]

Kinglets

Order: Passeriformes   Family: Regulidae

Ruby-crowned kinglet, Corthylio calendula

Golden-crowned kinglet, Regulus satrapa

Waxwings

Order: Passeriformes   Family: Bombycillidae

Bohemian waxwing, Bombycilla garrulus (casual or accidental)

Cedar waxwing, Bombycilla cedrorum (accidental)

Nuthatches

Order: Passeriformes   Family: Sittidae

Red-breasted nuthatch, Sitta canadensis (accidental)[2]

Wrens

Order: Passeriformes   Family: Troglodytidae

Pacific wren, Troglodytes pacificus

Mockingbirds and thrashers

Order: Passeriformes   Family: Mimidae

Northern mockingbird, Mimus polyglottos (accidental)[2]

Starlings

Order: Passeriformes   Family: Sturnidae

European starling, Sturnus vulgaris (accidental)

Dippers

Order: Passeriformes   Family: Cinclidae

American dipper, Cinclus mexicanus (resident*)

Thrushes and allies

Order: Passeriformes   Family: Turdidae

Mountain bluebird, Sialia currucoides (accidental)

Gray-cheeked thrush, (Catharus minimus)

Swainson's thrush (Catharus ustulatus (accidental)

Hermit thrush, Catharus guttatus (summer*)

Figure 35. Hermit thrush

This image is by Albuttlee at English Wikipedia, CC BY-SA 3.0, https://commons.wikimedia.org/w/index.php?curid=9045679; HermitThrush63 - Hermit thrush - Wikipedia. Permission is granted to copy, distribute and/or modify this document under the terms of the GNU Free Documentation License, Version 1.2 or any later version published by the Free Software Foundation.

Wood thrush, Hylocichla mustelina (accidental)

Eyebrowed thrush, Turdus obscurus (migrant)

Dusky thrush, Turdus eunomus (intermittent)

American robin, Turdus migratorius (casual)

Varied thrush, Ixoreus naevius

Old World flycatchers

Order: Passeriformes   Family: Muscicapidae

Gray-streaked flycatcher, Muscicapa griseisticta (intermittent)

Asian brown flycatcher, Muscicapa dauurica (casual or accidental)

Dark-sided flycatcher, Muscicapa sibirica (casual)

Siberian blue robin, Larvivora cyane (accidental)

Rufous-tailed robin, Larvivora sibilans (accidental), Cyanecula svecica

Siberian rubythroat, Calliope calliope (migrant)

Red-flanked bluetail, Tarsiger cyanurus (casual)

Narcissus flycatcher, Ficedula narcissina (casual or accidental)

Mugimaki flycatcher, Ficedula mugimaki (accidental)

Taiga flycatcher, Ficedula albicilla (intermittent)

Common redstart, Phoenicurus phoenicurus (accidental)

Northern wheatear, Oenanthe oenanthe (intermittent)

Accentors

Order: Passeriformes   Family: Prunellidae

Siberian accentor, Prunella montanella (accidental)

Wagtails and pipits

Order: Passeriformes   Family: Motacillidae

Eastern yellow wagtail, Motacilla tschutschensis (migrant)

Gray wagtail, Motacilla cinerea (intermittent)

White wagtail, Motacilla alba (migrant)

Olive-backed pipit, Anthus hodgsoni (intermittent or casual)

Pechora pipit, Anthus gustavi (casual)

Red-throated pipit, Anthus cervinus (intermittent)

American pipit, Anthus rubescens (summer*)

Finches, euphonias, and allies

Order: Passeriformes   Family: Fringillidae

Brambling, Fringilla montifringilla (migrant)

Hawfinch, Coccothraustes coccothraustes (accidental)

Common rosefinch, Carpodacus erythrinus (accidental)

Pallas's rosefinch, Carpodacus roseus (accidental)

Pine grosbeak, Pinicola enucleator (resident*)

Eurasian bullfinch, Pyrrhula pyrrhula (accidental)

Asian rosy-finch, Leucosticte tephrocotis (accidental)

Gray-crowned rosy-finch, Leucosticte tephrocotis (resident*)

Purple finch, Haemorhous purpureus (accidental)

Oriental greenfinch, Chloris sinica (accidental)

Common redpoll, Acanthis flammea

Hoary redpoll, Acanthis hornemanni

Red crossbill, Loxia curvirostra

White-winged crossbill, Loxia leucoptera

Eurasian siskin, Spinus spinus (accidental)

Pine siskin, Spinus pinus

Longspurs and snow buntings

Order: Passeriformes   Family: Calcariidae

Lapland longspur, Calcarius lapponicus (summer*)

Snow bunting, Plectrophenax nivalis (resident*)

McKay's bunting, Plectrophenax hyperboreus (casual or intermittent)

Old World buntings

Order: Passeriformes   Family: Emberizidae

Pine bunting, Emberiza leucocephalos (casual or accidental)

Little bunting, Emberiza pusilla (casual)

Rustic bunting, Emberiza rustica (intermittent)

Yellow-throated bunting, Emberiza elegans (accidental)

Yellow-breasted bunting, Emberiza aureola (casual)

Gray bunting, Emberiza variabilis (casual)

Pallas's bunting, Emberiza pallasi (accidental)

Reed bunting, Emberiza schoeniclus (casual)

New World sparrows

Order: Passeriformes   Family: Passerellidae

Chipping sparrow, Spizella passerina (accidental)

Fox sparrow, Passerella iliaca (summer*)

American tree sparrow, Spizelloides arborea (accidental)

Dark-eyed junco, Junco hyemalis (casual)

White-crowned sparrow, Zonotrichia leucophrys (accidental)

Golden-crowned sparrow, Zonotrichia atricapilla

Savannah sparrow, Passerculus sandwichensis (summer*)

Song sparrow, Melospiza melodia (resident*)

Lincoln's sparrow, Melospiza lincolnii (accidental)

Troupials and allies

Order: Passeriformes Family: Icteridae

Red-winged blackbird, Agelaius phoeniceus (accidental)

Brown-headed cowbird, Molothrus ater (accidental)

Rusty blackbird, Euphagus carolinus (accidental)

Brewer's blackbird, Euphagus cyanocephalus (accidental)

New World warblers

Order: Passeriformes Family: Parulidae

Ovenbird, Seiurus aurocapilla (accidental)

Northern waterthrush, Parkesia noveboracensis (accidental)

Tennessee warbler, Leiothlypis peregrina (accidental)

Orange-crowned warbler, Leiothlypis celata (casual)

Cape May warbler, Setophaga tigrina (accidental)

Magnolia warbler, Setophaga magnolia (accidental)

Yellow warbler, Setophaga petechia (summer*)

Blackpoll warbler, Setophaga striata (accidental)

Palm warbler, Setophaga palmarum (accidental)

Yellow-rumped warbler, Setophaga coronata (casual)

Townsend's warbler, Setophaga townsendi (accidental)

Wilson's warbler, Cardellina pusilla (summer*)

Cardinals and allies

Order: Passeriformes Family: Cardinalidae

Western tanager, Piranga ludoviciana (accidental)

Vagrant birds

Since the outermost of the Aleutian Islands, Attu, is relatively close to Far East Russia, it is not surprising that birds from the latter area sometimes occur on the Aleutians, usually driven there by inclement weather. Species include the Common rosefinch, Lanceolated warbler, and the first North American record of the Intermediate egret.

Figure 36. This image is of a male Cassin's Rosefinch, which is similar in appearance to the Common Rosefinch.

This image is by http://www.naturespicsonline.com/ - http://www.naturespicsonline.com/, Copyrighted free use, https://commons.wikimedia.org/w/index.php?curid=974171; en.wikipedia.org/wiki/Cassin%27s_finch#/media/File:Cassin's_Finch_(male).jpg. The copyright holder of this work allows anyone to use it for any purpose including unrestricted redistribution, commercial use, and modification.

Figure 37. Lanceolated warbler.

The Lanceolated warbler (Locustella lanceolata) is a species of Old World warbler in the genus Locustella. It breeds from northeast European Russia across the Palearctic to northern Hokkaidō, Japan. It is migratory, wintering in Southeast Asia. The genus name Locustella is from Latin and is a diminutive of locusta, "grasshopper". This refers to the song of the common grasshopper warbler and some others in this genus. The specific lanceolata is Latin for "spear-shaped" and refers to the streaks on the breast.

This figure is by Александр Чегодаев - https://www.inaturalist.org/photos/295726934, CC0, https://commons.wikimedia.org/w/index.php?curid=134572807; /en.wikipedia.org/wiki/Lanceolated_warbler#/media/File:Locustella_lanceolata_295726934.jpg. This file is made available under the Creative Commons CC0 1.0 Universal Public Domain Dedication.

Figure 38. Intermediate egret (aka Plumed egret)

This image is by John Gould. The image is in the public domain in countries where the copyright term is 100 years after the death of the author.

The Russian Aleutians

Although most of the Aleutian Islands belong to the USA, those in the Commander group of islands belong to Russia. They comprise Bering Island, Medny Island, Sea Lion Rock, Sea Otter Rocks, Tufted Puffin Rock (Kamen Toporkov or Ostrov Toporkov) and Kamen Ariy. It is hard to include an island called Tufted Puffni Rock and not show a photograph of its eponymous denizens.

Figures 39 – 40. A group of Tufted puffins - and a single Common Guillimot (murre): Can you spot it?

This image is by Judy Alderson/USFWS. - https://www.facebook.com/AlaskaMaritimeNationalWildlifeRefuge/photos/a.460515650636016/2044255845595314/?type=3&theater, Public Domain, https://commons.wikimedia.org/w/index.php?curid=73068302; /en.wikipedia.org/wiki/Tufted_puffin#/media/File:Group_of_Tufted_Puffins_(and_a_couple_of_Murres)_on_Bogoslof_Island_by_Judy_Alderson_USFWS.jpg. This work is in the public domain in the United States because it is a work prepared by an officer or employee of the United States Government as part of that person's official duties under the terms of Title 17, Chapter, Section 105 of the US Code.

This image is by Mary Sue Brancato/National Oceanic and Atmospheric Administration - https://sanctuaries.noaa.gov/about/history/ Time and Tide: A History of the National Marine Sanctuary System, p. 209, Public Domain, https://commons.wikimedia.org/w/index.php?curid=154328273; en.wikipedia.org/wiki/Tufted_puffin#/media/File:Olympic_Coast_National_Marine_Sanctuary_tufted_puffins.png. This image is in the public domain because it contains materials that originally came from the U.S. National Oceanic and Atmospheric Administration, taken or made as part of an employee's official duties.

3. Conclusions

How great to spend a month on Attu. However, no transport goes there at present. Maybe I'll have to be content with a daytrip to Dutch Harbor.

Made in the USA
Monee, IL
05 November 2026

67985536R00038